TRACING NUMBERS

PRE-SCHOOLERS PRACTICE WRITING NUMBERS WORKBOOK, KIDS AGES 3-5

BOOK 2

By Brighter Hand

http://sudokuprintable.blogspot.com

COPYRIGHT NOTICE

Copyright © 2017 by **TRACING NUMBERS PRESCHOOLERS PRACTICE WRITING NUMBERS WORKBOOK, KIDS AGES 3-5**.

All rights reserved. This book or any portion thereof may not be reproduced or used in any manner whatsoever without the express written permission of the publisher except for the use of brief quotations in a book review.

www.ingramcontent.com/pod-product-compliance
Lightning Source LLC
Chambersburg PA
CBHW082221220526
45470CB00010B/3260

FIND OUT MORE HERE:

http://sudokuprintable.blogspot.com

ALSO BE INTERESTED IN MY NEWEST COLLECTION OF BOOKS

CHECK OUT THE BOOKS ON THE NEXT PAGE

RECOMMEND ALSO:

***SUDOKU:500 Sudoku Puzzles

by Champ Lopez

***Tracing Letter Pre-schoolers Practice Writing ABC Alphabet Workbook, Kids Age 3-5 by Brighter Hand

***Times Ultimate Mind Games Book Killer Su doku Over 300 Puzzles Book 1

by Champ Lopez

***Lots of Fun Number Tracing Practice Learn Numbers 0 to 20

by Handwriting Time

CONTENTS

INTRODUCTION	1
HOW TO USE THIS BOOK	2
SMALL NUMBER WORDS EXERCISE 1	*3*
SMALL NUMBER WORDS EXERCISE 2	*4*
SMALL NUMBER WORDS EXERCISE 3	*5*
SMALL NUMBER WORDS EXERCISE 4	*6*
SMALL NUMBER WORDS EXERCISE 5	*7*
SMALL NUMBER WORDS EXERCISE 6	*8*
SMALL NUMBER WORDS EXERCISE 7	*9*
SMALL NUMBER WORDS EXERCISE 8	*10*
SMALL NUMBER WORDS EXERCISE 9	*11*
SMALL NUMBER WORDS EXERCISE 10	*12*
SMALL NUMBER WORDS EXERCISE 11	*13*
SMALL NUMBER WORDS EXERCISE 12	*14*
SMALL NUMBER WORDS EXERCISE 13	*15*
SMALL NUMBER WORDS EXERCISE 14	*16*
SMALL NUMBER WORDS EXERCISE 15	*17*
SMALL NUMBER WORDS EXERCISE 16	*18*
SMALL NUMBER WORDS EXERCISE 17	*19*
SMALL NUMBER WORDS EXERCISE 18	*20*
SMALL NUMBER WORDS EXERCISE 19	*20*
SMALL NUMBER WORDS EXERCISE 20	*22*
SMALL NUMBER WORDS EXERCISE 21	*23*
SMALL NUMBER WORDS EXERCISE 22	*24*
SMALL NUMBER WORDS EXERCISE 23	*25*
SMALL NUMBER WORDS EXERCISE 24	*26*
SMALL NUMBER WORDS EXERCISE 25	*27*
SMALL NUMBER WORDS EXERCISE 26	*28*
SMALL NUMBER WORDS EXERCISE 27	*29*
SMALL NUMBER WORDS EXERCISE 28	*30*
SMALL NUMBER WORDS EXERCISE 29	*31*
SMALL NUMBER WORDS EXERCISE 30	*32*
SMALL NUMBER WORDS EXERCISE 31	*33*
SMALL NUMBER WORDS EXERCISE 32	*34*

SMALL NUMBER WORDS EXERCISE 33 *35*
SMALL NUMBER WORDS EXERCISE 34 *36*
SMALL NUMBER WORDS EXERCISE 35 *37*
SMALL NUMBER WORDS EXERCISE 36 *38*
SMALL NUMBER WORDS EXERCISE 37 *39*

INTRODUCTION

A child who learns to trace letters, numbers at home, at the early age, with their loving parent or caregiver, grows in self-confidence and independence. This promotes greater maturity, increases discipline and lays the basis for moral literacy. A child who begins with early learning books has a distinct advantage over his or her peers. One of the big advantages being there is no psychological pressure.

HOW TO USE THIS BOOK

Practice, Practice, Practice makes life easier and worthwhile so train your child analytical mind by tracing letters in the alphabet the conventional way through handwritings as the saying said: **"Young children need writing to help them learn about reading, they need reading to help them learn about writing; and they need oral language to help them learn about both."**

SMALL NUMBER WORDS EXERCISE 1

eleven

sixty-three

ninety-two

twenty-eight

fifty-five

two

SMALL NUMBER WORDS EXERCISE 2

seven

sixty five

seventy four

eighty one

sixty seven

fifty six

SMALL NUMBER WORDS EXERCISE 3

sixty-two

sixty-four

forty-two

ninety

fifty-four

eighteen

SMALL NUMBER WORDS EXERCISE 4

forty three

sixty eight

sixteen

fifty seven

sixty

forty two

SMALL NUMBER WORDS EXERCISE 5

twenty three

ten

ninety six

seventy two

fifty

seventy five

SMALL NUMBER WORDS EXERCISE 6

forty

ninety-two

ninety-five

fifty

five

sixty-three

SMALL NUMBER WORDS EXERCISE 7

sixty eight

fifty one

fifteen

five

seventy one

thirty six

SMALL NUMBER WORDS EXERCISE 8

fifty nine

thirty two

five

forty eight

nine

ninety five

SMALL NUMBER WORDS EXERCISE 9

ninety eight

thirty eight

thirty

seven

thirty nine

seventy five

SMALL NUMBER WORDS EXERCISE
10

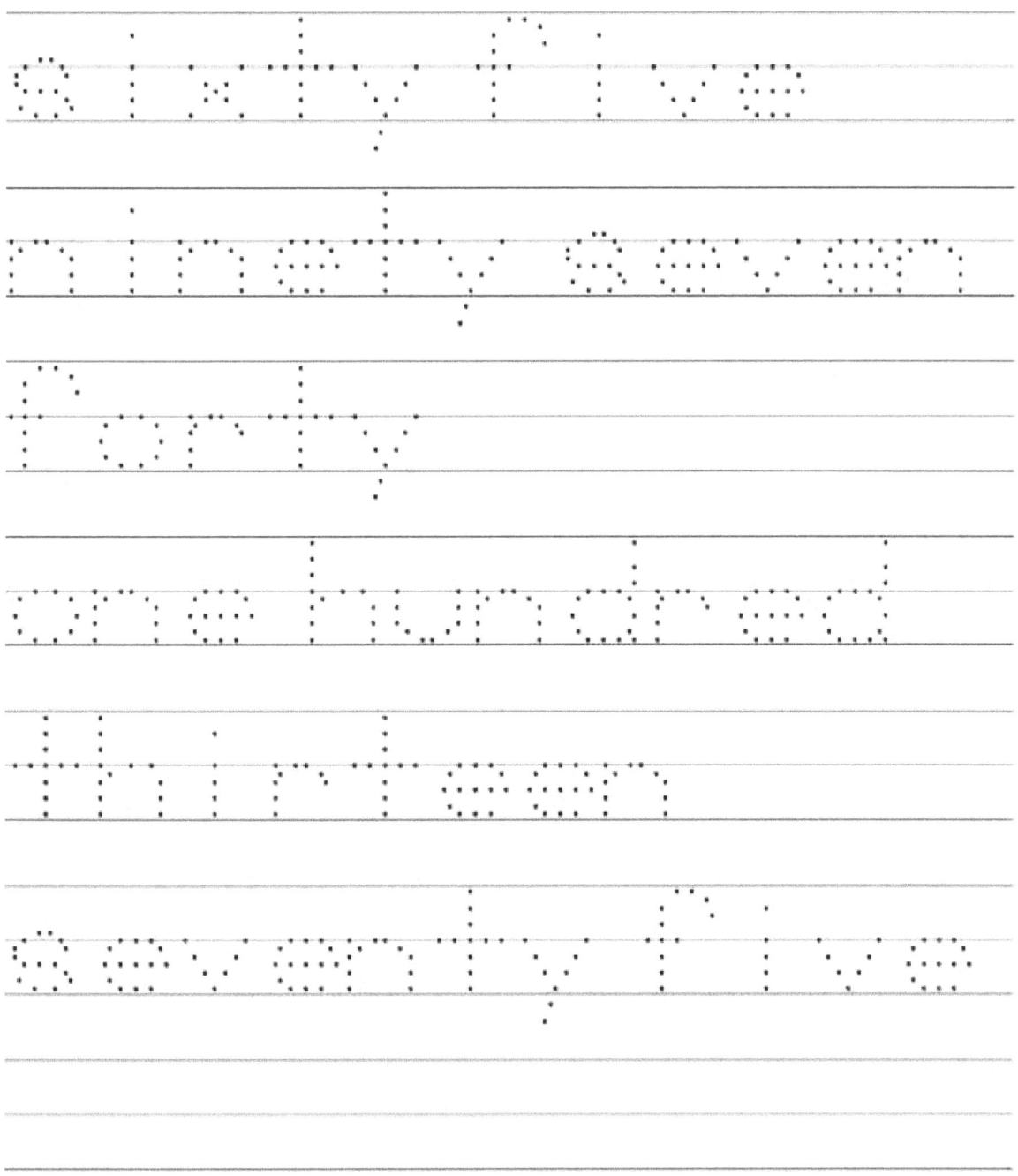

sixty five

ninety seven

forty

one hundred

thirteen

seventy five

SMALL NUMBER WORDS EXERCISE
11

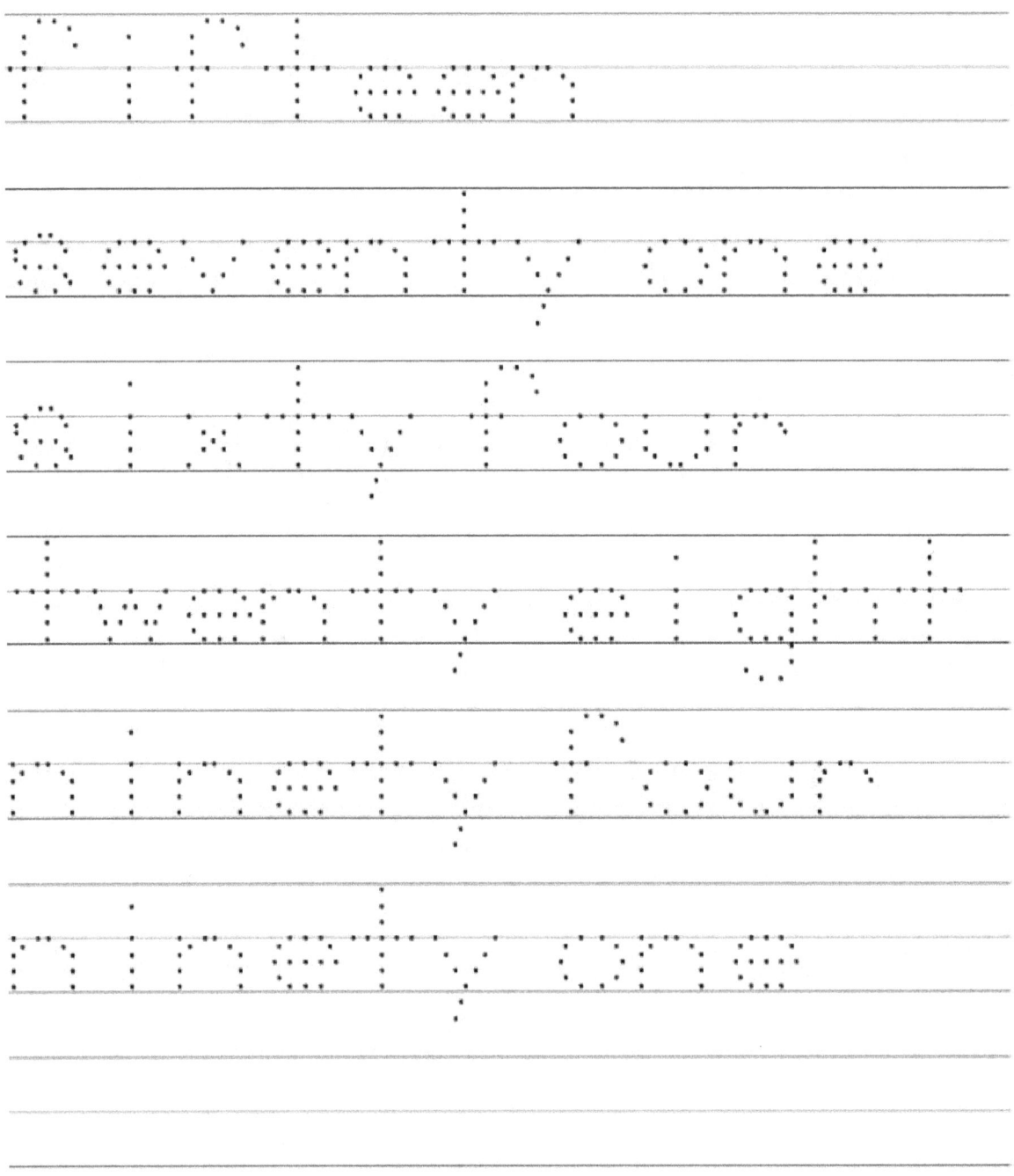

fifteen

seventy one

sixty four

twenty eight

ninety four

ninety one

SMALL NUMBER WORDS EXERCISE
12

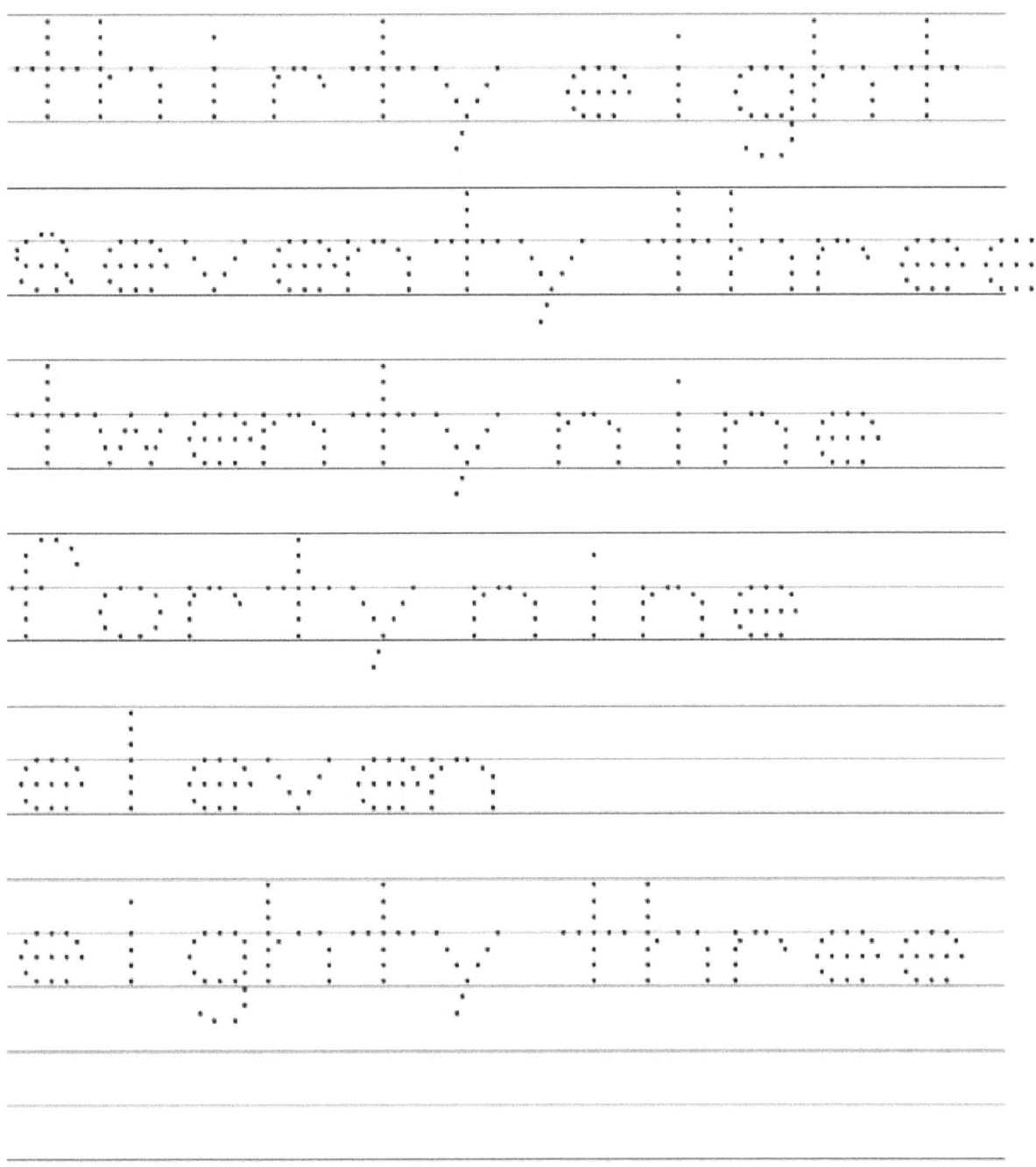

thirty eight

seventy three

twenty nine

forty nine

eleven

eighty three

SMALL NUMBER WORDS EXERCISE
13

twenty eight

two

fourteen

seventy four

sixty three

fifty five

SMALL NUMBER WORDS EXERCISE
14

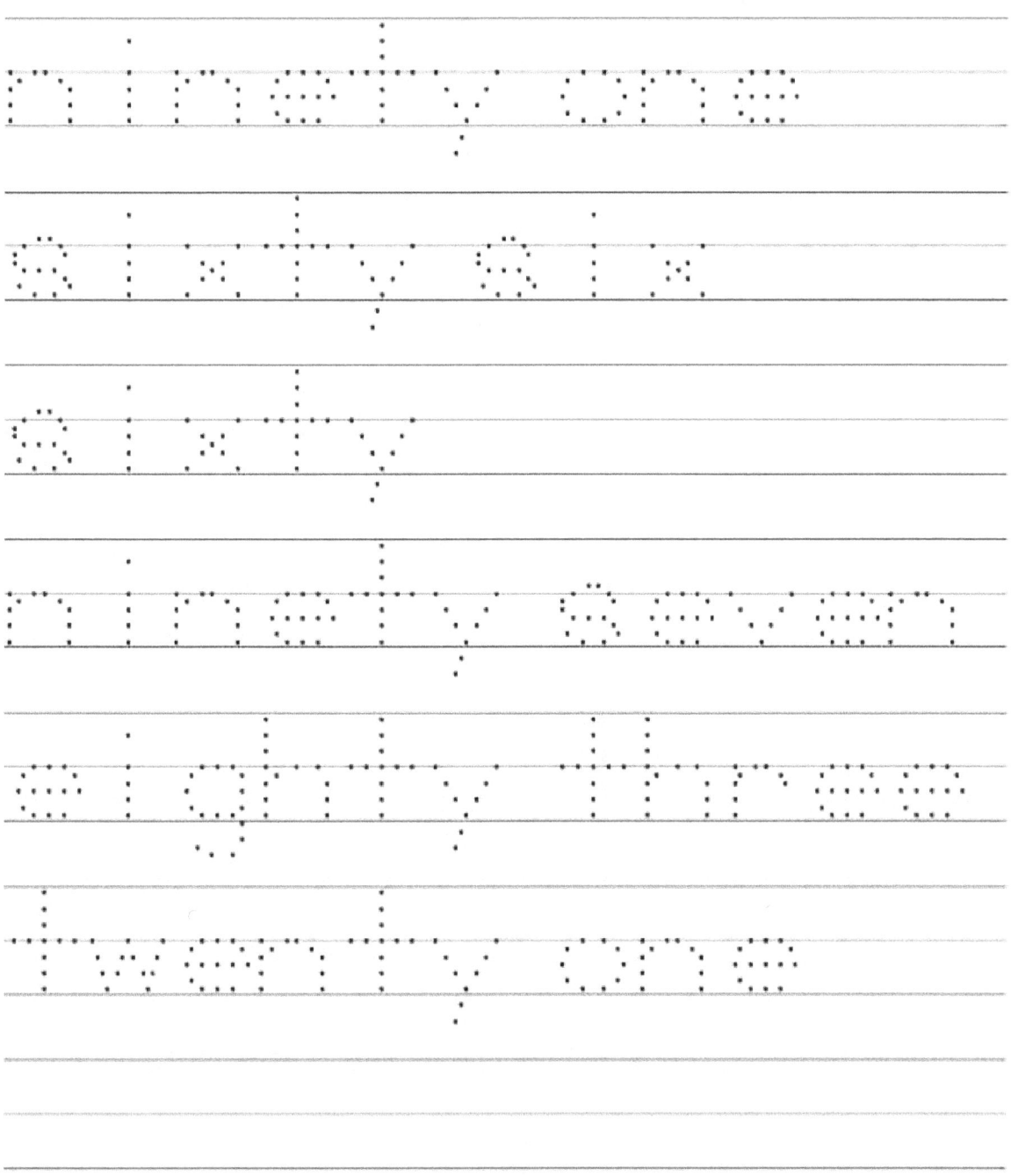

ninety one

sixty six

sixty

ninety seven

eighty three

twenty one

SMALL NUMBER WORDS EXERCISE
15

eighteen

thirty five

forty eight

ten

forty three

ninety three

SMALL NUMBER WORDS EXERCISE
16

twenty four

ninety six

seven

seven

forty two

fifty two

SMALL NUMBER WORDS EXERCISE 17

sixty one

thirty one

sixty five

seventy six

twenty seven

eleven

SMALL NUMBER WORDS EXERCISE
18

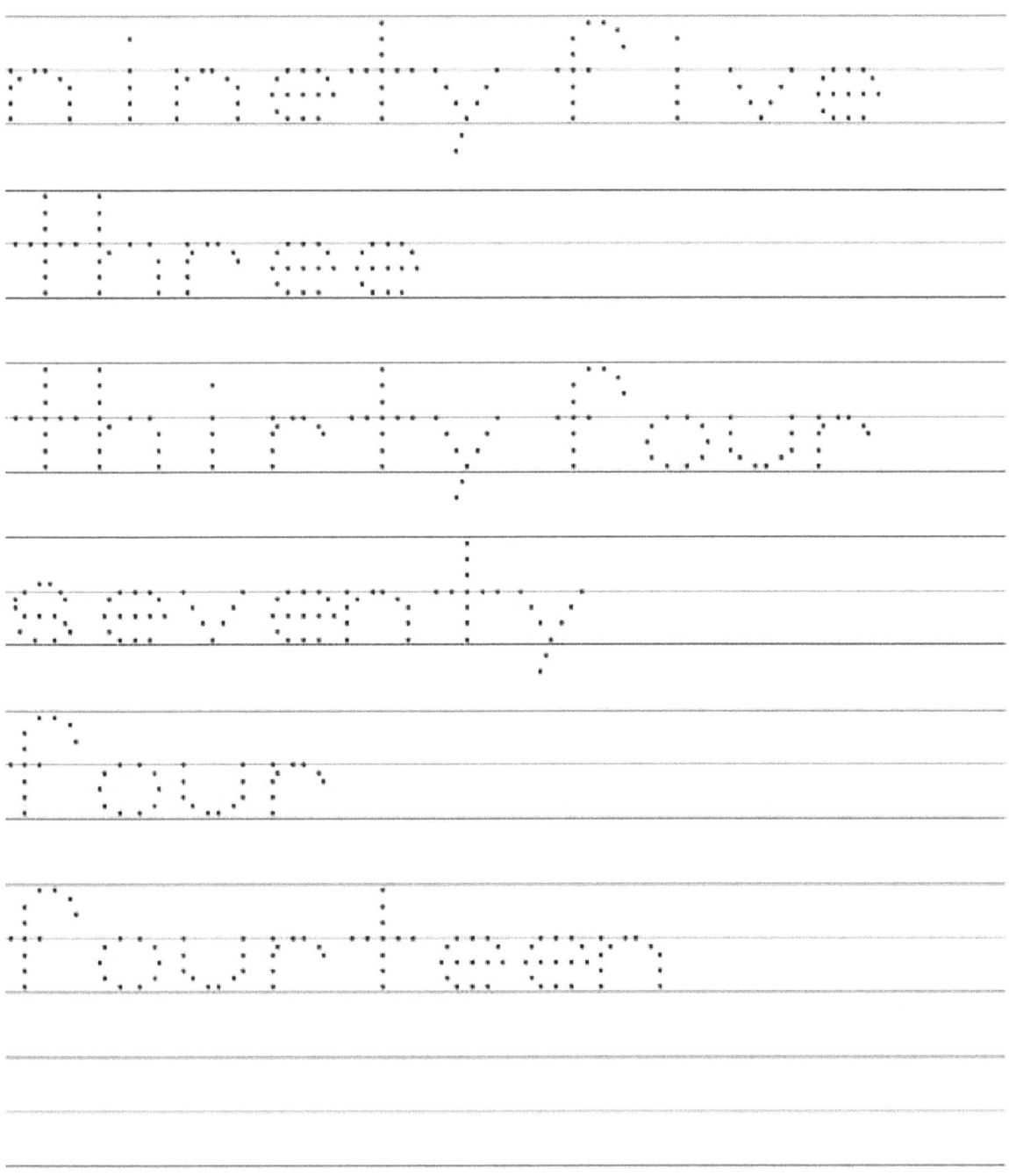

ninety five

three

thirty four

seventy

four

fourteen

SMALL NUMBER WORDS EXERCISE
19

twenty nine

nineteen

eighty five

fourteen

thirty two

ninety four

SMALL NUMBER WORDS EXERCISE
20

forty four

sixty six

six

eighty nine

twenty

thirty seven

SMALL NUMBER WORDS EXERCISE
21

eighty eight

twenty one

nineteen

twenty seven

eighty nine

ninety two

SMALL NUMBER WORDS EXERCISE
22

ten

fifty

sixty-three

eleven

sixty-seven

four

SMALL NUMBER WORDS EXERCISE 23

sixty five

sixty eight

forty four

thirty one

ninety four

sixty seven

SMALL NUMBER WORDS EXERCISE 24

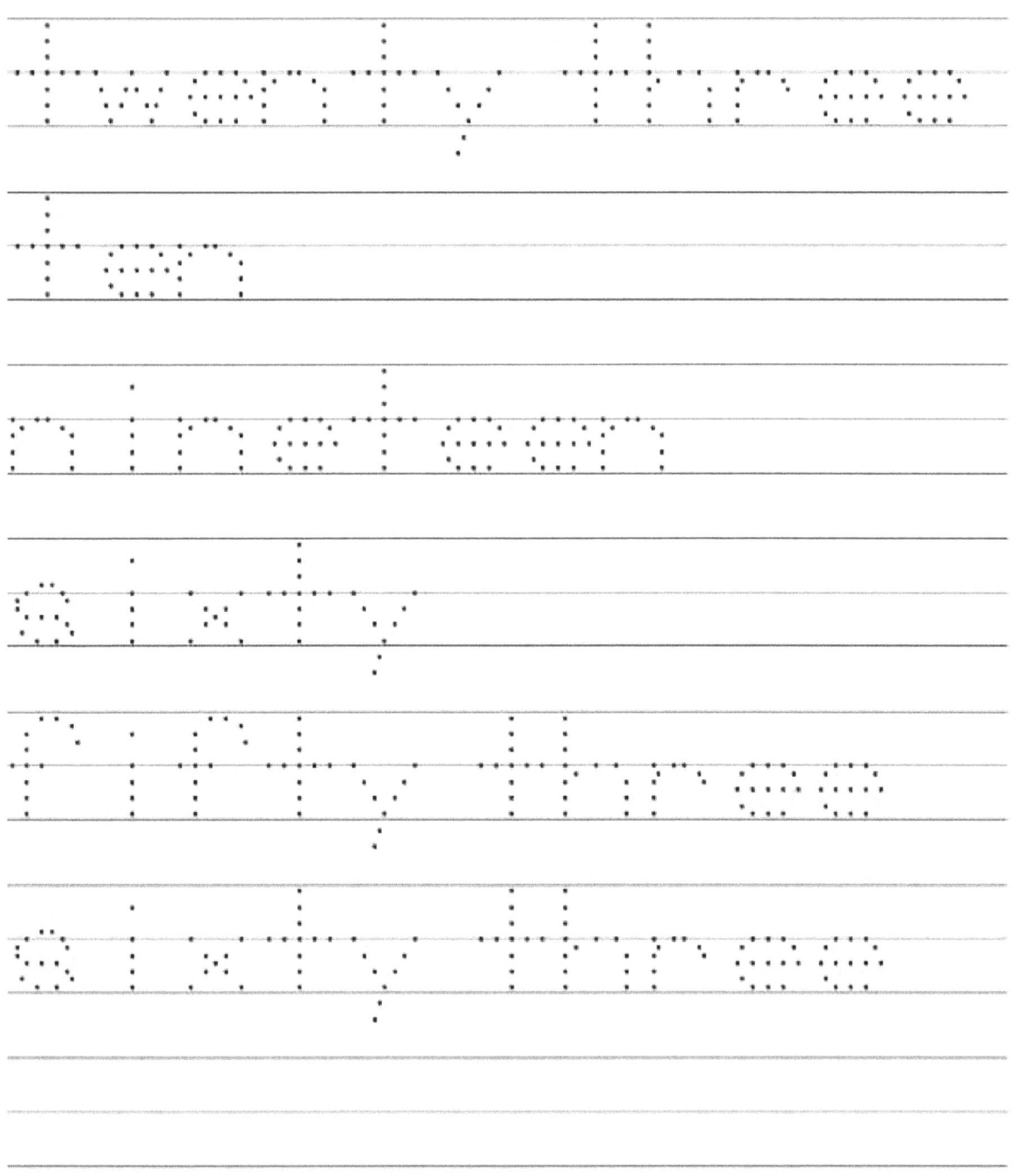

twenty three

ten

nineteen

sixty

fifty three

sixty three

SMALL NUMBER WORDS EXERCISE
25

four

sixty seven

fifty two

ten

seventy three

eighty two

SMALL NUMBER WORDS EXERCISE
26

sixty five

thirty

twenty nine

thirty seven

eighty

eighty two

SMALL NUMBER WORDS EXERCISE
27

ninety

fifty nine

seventy eight

four

thirty eight

fourteen

SMALL NUMBER WORDS EXERCISE
28

ninety

fifty nine

seventy eight

four

thirty eight

fourteen

SMALL NUMBER WORDS EXERCISE
29

forty five

eighty

fifty one

ninety four

six

seventy two

SMALL NUMBER WORDS EXERCISE
30

Fifty six

four

Sixty seven

eighty five

fifty

fifty four

SMALL NUMBER WORDS EXERCISE
31

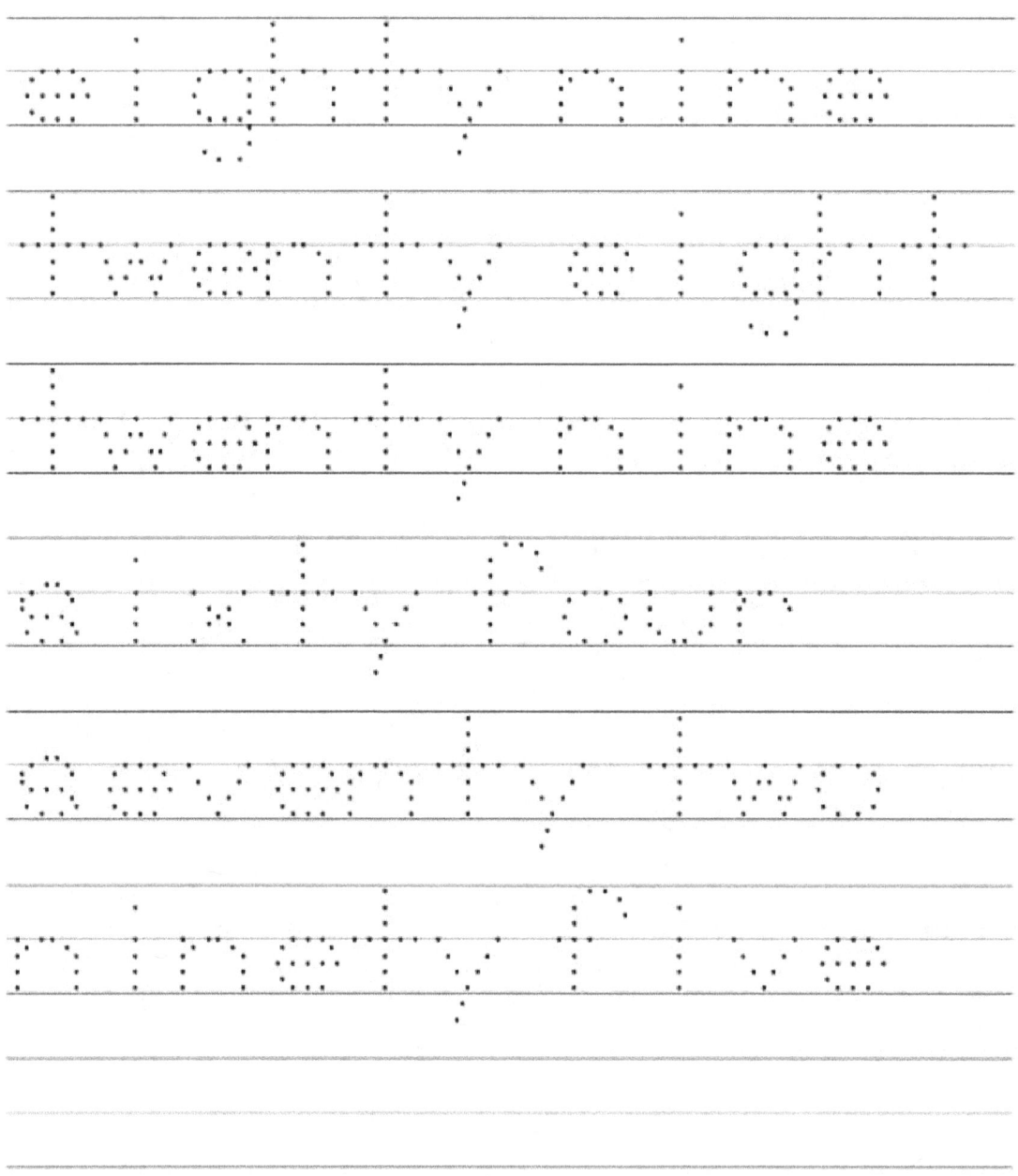

eighty nine
twenty eight
twenty nine
sixty four
seventy two
ninety five

SMALL NUMBER WORDS EXERCISE 32

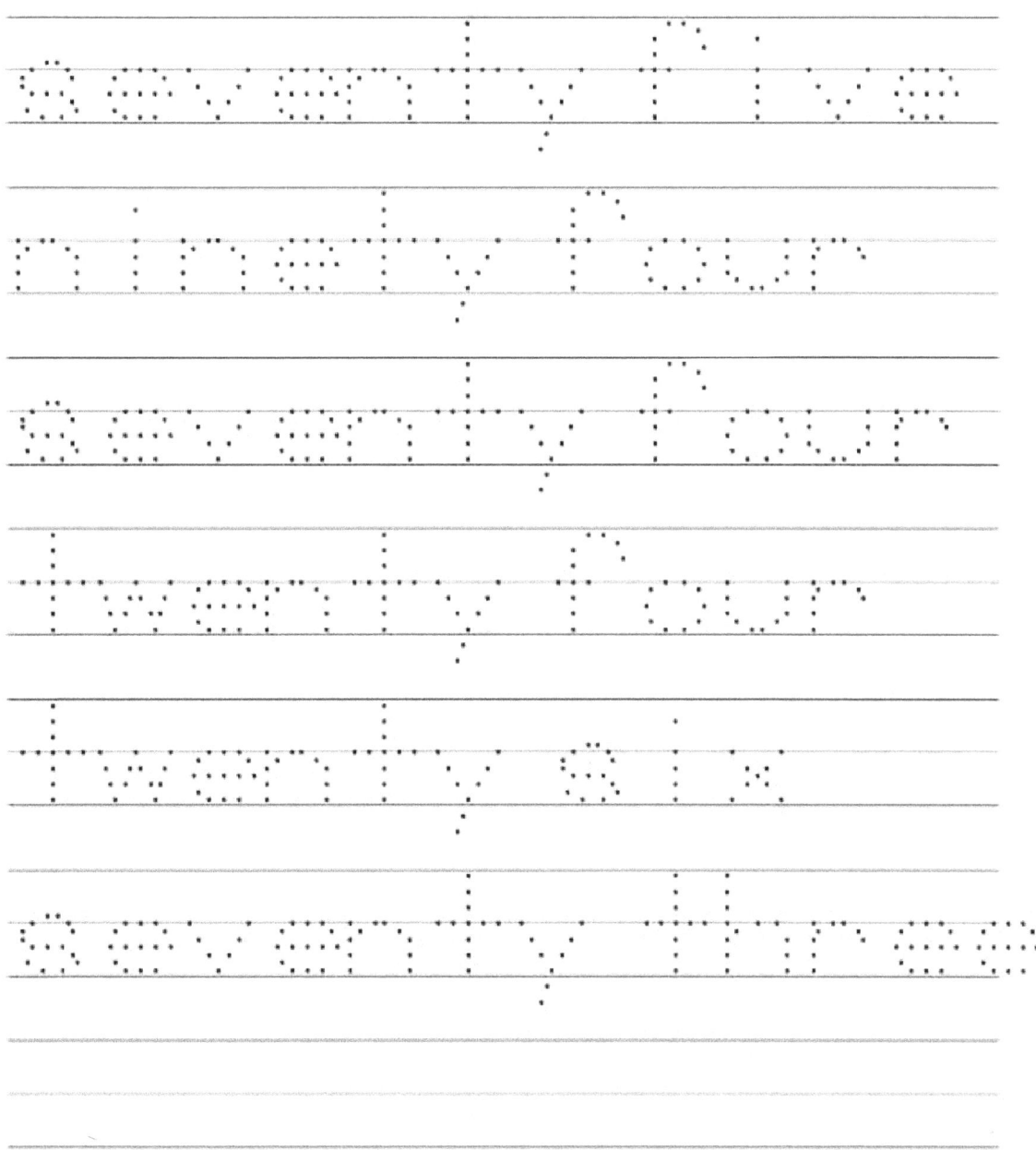

seventy five
ninety four
seventy four
twenty four
twenty six
seventy three

SMALL NUMBER WORDS EXERCISE
33

seventeen

thirty

five

seventeen

ninety two

thirty one

SMALL NUMBER WORDS EXERCISE
34

thirty

eighty six

two

eight

twenty eight

sixty two

SMALL NUMBER WORDS EXERCISE 35

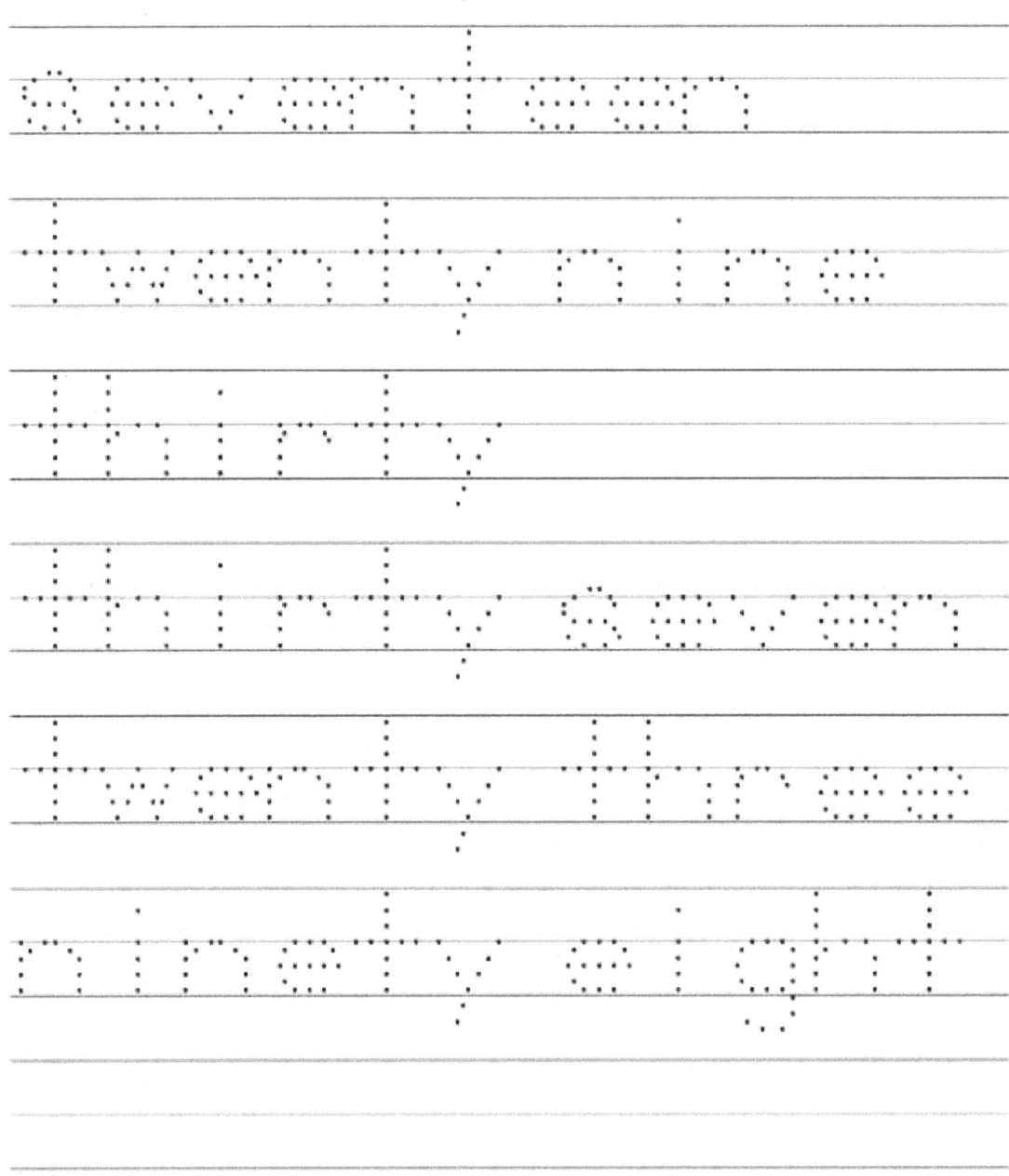

seventeen
twenty nine
thirty
thirty seven
twenty three
ninety eight

SMALL NUMBER WORDS EXERCISE 36

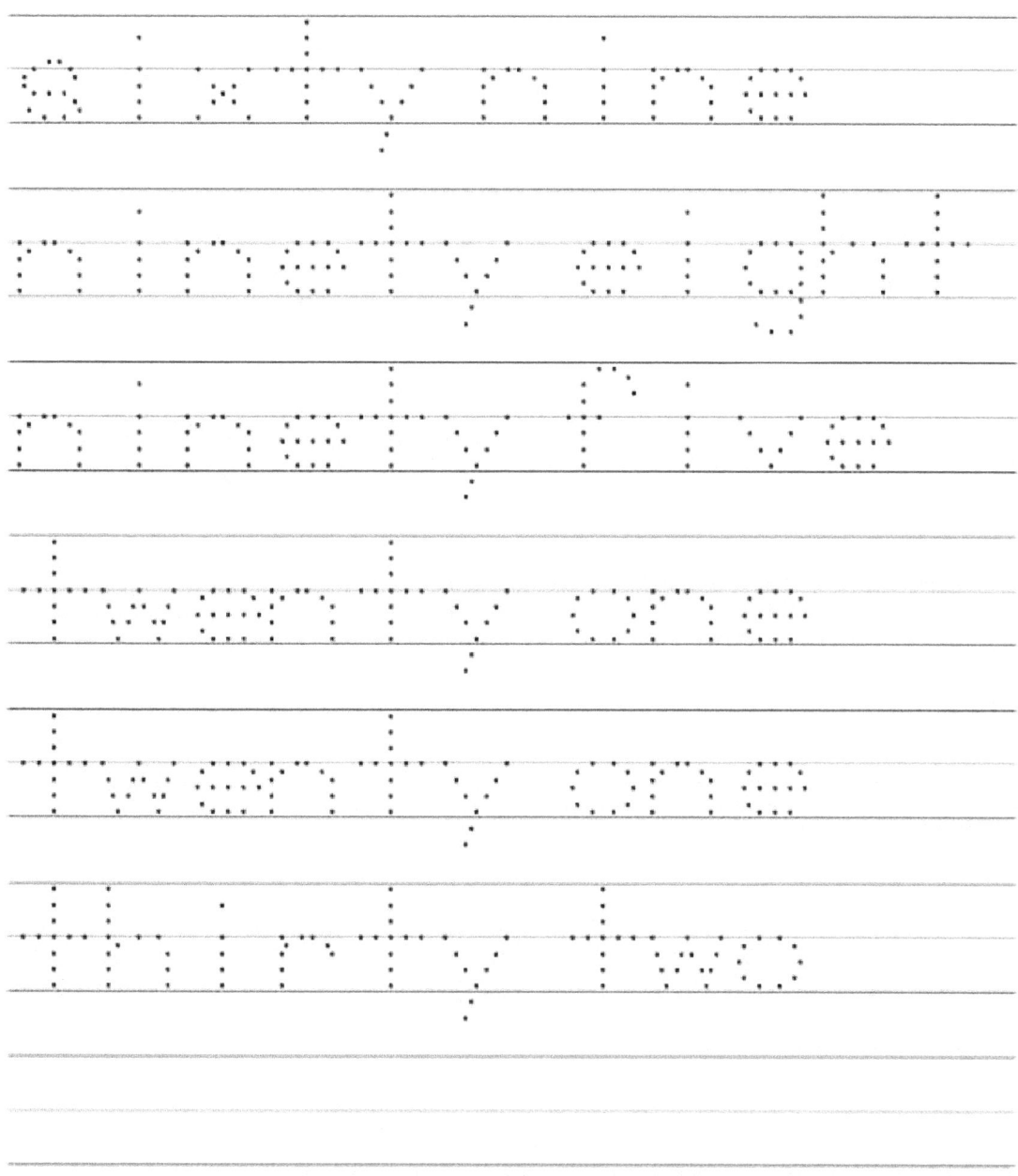

sixty nine

ninety eight

ninety five

twenty one

twenty one

thirty two

SMALL NUMBER WORDS EXERCISE 37

three

twenty eight

eighty two

sixty three

nine

ninety two